What Color Is Your Cat?

A Personality Guide for Every Shade of Feline

by Cathy Crimmins

Sourcebooks Hysteria™
An Imprint of Sourcebooks, Inc.®
Naperville, Illinois

Cover photo courtesy of Corbis
Internal photos on pages viii, 32, 41, 45, 57, 67, 76 courtesy of Corbis
Internal photos on pages 5, 8, 17, 20, 46, 51, 58, 68, 83, 87 courtesy of Sourcebooks, Inc.
Sourcebooks would like to thank Mimi, Jake, Snickers, Casper, Rocky, Pepper, Heidi, and Genghis

Published by Sourcebooks, Inc.
P.O. Box 4410, Naperville, Illinois 60567-4410
(630) 961-3900
FAX: (630) 961-2168
www.sourcebooks.com

Library of Congress Cataloging-in-Publication Data

Crimmins, C.E.
What color is your cat?: a personality guide for every shade of feline/Cathy Crimmins.
p.cm.
Includes bibliographical references (p.).
ISBN 1-887166-74-2 (alk. paper)
Cats—Behavior. 2. Cats—Color. I. Title.

SF446.5 .C75 2001
636.8'0887—dc21

2001031321

Printed and bound in China
IM 10 9 8 7 6 5 4 3 2 1

Table of Contents

Acknowledgments

Thanks to my own current kitties, Lemon Leo, Cosmo, and Daisy, for their continued inspiration, and to all the dear departed felines in my life. I hated losing them, but their memories are enshrined in this book. Also, thanks to everyone at Sourcebooks, especially Deb Werksman. Many people either shared stories or showed patience, including Anne Kaier, Joellen Brown, Alan Forman, Kelly Crimmins, Bette Lancaster, Sarah Babaian, Tom Maeder, Aishe Ozbekhan, Betty Crane, Elizabeth McKinstry, Barbara Patterson, Gretchen Worden, and Betsy Cooper.

Introduction

I love cats, especially alley cats. I actually have a bonafide alley behind my house. In my neighborhood, cats seem to rain from the sky. That's why I've owned and "out-placed" so many strays. And that's why I've been able to notice, over the years, that ordinary cats of different colors have distinctly different personality traits. I've come to be able to predict whether a cat will bite, whether it will sit in a human lap, or whether it will stick around to greet dinner guests, all on the basis of its color and patterning. For example, the longer a cat's hair, the lower its I.Q. An orange cat will never sit on your lap; gray cats are prone to obesity; tabbies are the most dog-like; and calico cats top the list for most neurotic.

While these generalizations might seem prejudiced, I like to think that being able to track these personalities means what those of us who love domestic short hair cats have always known: even alley cats are thoroughbreds! I've written this book for everyone who has ever loved, lived with, been nurtured by, and been driven crazy by a cat. What color is your cat?

What Color Is Your Cat?

A kitten is a rosebud in the garden
of the animal kingdom.
—Robert Southey

A Very Short "Course" in Cat Genetics

We marvel at the many variations of *Felis catus*, the domestic cat—different colors, different patterns, different sizes. The alley cat population is very, shall we say, promiscuous (the scientific term is heterozygous), meaning that its members intermingle with a wide range of sex partners, introducing the diversity that makes for a strong species. (Did you ever wonder why plain ordinary cats are so much healthier than purebred types? The answer could lie in their genes!)

Since more than one male might be fertilizing the eggs of the female, you might see many varieties of kittens, even in one litter.

When trying to understand cat genes, as with any genes, there are really only two principles—genes are either dominant or recessive. Genes come in pairs. A cat or a human gets one from the mother, and one from the father. A dominant gene will override the other gene in its pair to create its distinguishing characteristic. Take, for example, the human genes responsible for eye color. The gene for brown eyes is always dominant, and the one for blue is recessive. For a person to have blue eyes, he or she must have inherited the recessive gene for blue eyes from both parents.

Many pairs of dominant and recessive genes go into the equation of making the perfect alley cat. Stripes, spots, patches? Long or short hair? Much of it is the luck of the genetic draw, and your cat will inherit not only characteristics that are visible in his or her parents, but also hidden, recessive traits that will emerge as dominant when they pair up with other recessive genes.

Color

There are only three true colors in cats: black, red (orange), and white. These three colors are dominant, and if a cat happens to get the same color gene from each parent, then you have a strongly colored monochrome cat. Although, orange cats always have a bit of tabby patterning—I'll explain that later.

Other colors in the alley cat spectrum—brown, yellow, gray—are the result of a recessive gene that they call the dilute gene—that is, it waters down a dominant color. Think of the dilute gene as being a sort of whitewash that changes the stronger color. So, a cat carrying a dominant black gene with a recessive dilute gene, for example, will be—surprise!—gray. Another type of dilute black gene, when it is recessive, will create a chocolate colored cat instead of a black one. A dilute gene paired with a dominant red gene will produce a beautiful cream-colored feline.

Some of the colors are gender related—they occur predominantly in one sex or another. You will rarely find a female orange cat or a male calico. (For example, the percentage of calico, or tri-colored cats, that are female is 99.999 percent.)

Patterning

All cats are essentially tabby cats, because along with genetic markers for color, they possess what is known as the agoute gene—a set of directions that makes those beautiful stripes, and what some cat experts call mackereling, or a pattern of spots. But some colors can mask the tabby patterns. (Look closely at all kittens, even solid colored ones, and you will see a vague tabby pattern, even if they grow out of it later.) Only the red color cannot override the striped tabby gene, which is why all red or orange cats show some stripes.

The prevalence of the patterning gene explains why there are so many tabby cats around. I always think of the subtly striped black and gray tabby as the basic model. The tabby is the Volvo of cats.

Patching

Have you ever noticed how cats that seem one color always have a tiny patch of white somewhere? And how about those little white boots or chin goaties sported by some felines? Well, there is a modifying gene for creating white feet markings, which explains how so many alley cats get the name Boots. There is also a dominant marker for white locketing, those small, white areas on a cat's chest or groin.

Hair Length

Like all other traits, hair length is genetically encoded. Long hair is recessive, which explains why a smooth-coated cat can sometimes produce one or two longhaired kittens; she mated with another cat with the recessive gene for long hair. My vet believes that longhaired alley cats are mellower than the short hairs. I also think that they are not as bright as their smooth brothers and sisters. I love longhairs, but I've never met a brilliant one. I sometimes wonder if a stupidity gene is attached to the hair-length gene!

And, in case you are wondering what my ultimate kitty intelligent test is, I'm happy to reveal it here: the suitcase situation. My smartest cats have always turned weird as soon as I take out a suitcase. They know they will be abandoned, and many of them have refused even so much as to look at me once I unzip my valise. But my truly stupid cats have no inkling of cause and effect. One of them, Monty, a gloriously handsome longhaired chocolate alley cat, would jump in and out of the suitcase, playing as I packed. He also never understood that the appearance of a cat carrier meant he was going to the vet. While my other cats would have long disappeared down the stairs or under the bed at the first sight of the box, Monty would stand around blankly and then act amazed when I lifted him into it.

An Alley Cat Alphabet

A is for alley cats, all types and cute.
B is for black cats, with little white boots.
C is for calicos and beautiful fur,
D is for dappled pelts rubbed as they purr.
E is for eating—the dry or wet food,
F is for fat cats, the best of the brood.
G is for gangs of cats out on the town.
H is for hissy fits, fights of reknown.
I is for intellect—in cats it is firm!
J is for Jellicle, Tom Eliot's term,
K is for kittens, fuzzy and small.
L is for litter box, useful to all.

M is for meow, the universal greeting.
N is for napping, wherever there's seating.
O is for orange cats with cute tiger stripes,
P is for purring and sandpaper swipes.
Q is for queens, regal on chairs.
R is for rough coats, to the tips of their hairs.
S is for short hairs, the classic alley cat.
T is for tabby, the cutest spoiled brat.
U is for under—the chin, spot to stroke.
V is for vanity, a kitty cat's cloak.
W is for whiskers, the adornment of face.
X is for X-ray, stares into space.
Y is for yowling—Hey, have we met?
Z is for zanies—cats, our best pet!

Orange Cats

Most cats are okay, but orange ones are great.

—Harry Annally, from the core beliefs of
The Orange Cat Superiority Organization

Orange cats at-a-glance

Size: Big-boned
Tendency toward obesity (on a scale of 1 - 10): 7
Sex: Almost always male
Color combos: Reddish-orange solid, orange and white striped, beige
Peculiar traits: Like to get under foot, tendency to nibble on their owners, able to stare into space, Zen-like, for a very long time
Craziness (on a scale of 1 - 10): 8
Famous orange cats: Tonto of *Harry and Tonto,* Morris of Nine Lives, Topcat, Tony the Tiger, Heathcliff, Orion of *Men in Black*

Orange Cats: Our Domestic Tigers

Orange cats always seem a little dotty and a bit British, perhaps because of all the wonderful literary descriptions of marmalade or ginger cats. Like tabbies, they are long-lived. Once in a while you will see a frail fine-boned specimen, but usually the orange cat genes produce big-bodied tomcats. It's not unusual at all for a tiger cat to weigh over twenty pounds.

If you want a lap cat, don't get an orange. Sure, they like human company, and will sit at the end of your bed, next to you on the couch, or even beside your leg when you are in a straight chair. But, they hate being cuddled in a lap, and will not tolerate it for more than a minute. A friend's orange cannot be coaxed into her lap for anything, yet he will push the whole force of his twenty-five pounds against her shoulder in bed. My big orange cat sits like a rock at the end of the mattress all night.

Orange cats like to wander and often get lost. They have an independent spirit that can't be contained by doors or yards. My friend Anne adopted her orange cat, Henry, after finding him nearly dead by the side of the road, the victim of a car accident. After having him patched up at the vet's, it took her a year to coax him out from under her bed. But now she can't keep him inside. He roams all over the yard of her little city townhouse and into all the other neighbor's yards, too. He eats birds indiscriminately, much to the chagrin of the neighborhood birdwatchers.

Another orange cat trait: they have two speeds—stop, and *go*! They can stare straight into space for hours, sometimes lifting one paw as if in a yoga pose, or they are tearing around the house.

Orange cats are not very vocal, and don't purr easily. They do like to let loose with the occasional tiger-like growl. They love to roughhouse—their idea of showing affection, after you pet them a while, is to give you a few playful nips. All orange cats like to bite, but in varying degrees.

Orange cats do not get out of the way. You might feel as though yours is constantly trying to trip you. In a driveway, an orange cat will sometimes try to stare down an approaching car. Going down a flight of stairs, I actually put my foot on top of an orange cat and pretended to step down, and he *still* did not move.

Perhaps it is just the coloring, but marmalade cats do have certain lion-like dignity. They wait to be approached. Despite my affection for oranges, I didn't own one until I was in my thirties. And he was pre-owned! But even if you get an old orange cat, you'll get mileage out of him—my Lemon Leo is now over twenty and still going strong!

Orange Cats I Have Known and Loved

Lemon Leo: the Double Agent

Lemon Leo had passed through most of his nine lives before showing up on our doorstep during a blizzard, cut up and weighing almost nothing. The vet told me he was probably already eleven years old when we found him. Like all orange cats, he was stubborn and proud, and I finally had to drag him into the house to keep him from freezing to death. My little girl, then five, said he looked like a big lemon. My husband had always wanted a cat named Leo. And so Lemon Leo received his odd moniker.

After living several years with us, Lemon Leo began disappearing frequently, spending only a night or two per week at home. One evening he came back wearing a pink collar with rhinestones. A while after that, I found a note attached to the collar. "Is this your cat? We are worried that she is not being taken care of, and we are moving at the end of the week. Please call us to let us know that she will be okay."

Dialing the phone number provided, I got the dish on the double life the Lemster had been leading. The neighbors, two college students, had called "her" Big Red and fed her choice canned food. After they left, Lemon Leo went back to boring, dry food and a life as a macho, aging tomcat. We haven't mentioned his gender experimentation again.

To me, Lemon Leo is the perfect specimen of orange cathood. Although he is old, he is still very playful—he cadges the rubber bands from the table after we unroll the newspapers, and carries them all around the house, shooting them up into the air and chasing them as if they were very slim birds. He periodically has great bursts of energy in which he races up and down the stairs. He seems to have no perception of himself as an old large cat. Only recently I found him squeezing his big-boned body into a tiny shoebox. And even though he is over twenty, he can still knock over Christmas ornaments with the best of them.

Rufus: Two-Lifer

After years of not owning a cat, my mom adopted a gorgeous, orange kitty, Rufus, who is, in her estimation, the perfect pet. He was found wild as a kitten, and for the first three years he didn't come around much or even try to sit on the furniture. Still, my mother liked knowing he was there. She had a little cat door in her basement where he went in and out, and one great aspect of his personality was that he never used an indoor litter box.

Gradually, Rufus became friendlier, both to my mother and to other felines. He became a party animal who would invite other cats home to hang out in his house and eat his food. My mother would come home in the evening and see a strange Tomcat sitting in her living room. Yet she really grew to love Rufus, despite his misguided generosity towards strays and his propensity for leaving squirrel tails on her outdoor patio.

My mom travels a lot, and always has a neighbor look in on Rufus when she is gone. She was just getting home from a big trip, checking her messages, when she heard a terrible one from her vet—Rufus had been run over, found dead by a neighbor. What did she want to do? Shaking, she called the vet and said that he could cremate Rufus. Then she got off the phone and cried. Later that day, she called me to say, "Rufus is dead! He was such a good cat!"

Several hours later, Rufus walked into her living room. She began screaming, and he ran away. Then, of course, she recovered herself, and he came back up the cellar stairs to rub against her legs. "It was like seeing a ghost!" she told me. She's never found out who the other unfortunate orange cat was, but ever since that day she's felt as if Rufus is on his second life.

Vinnie: Blubber Boy

Vinnie, who is so fat that my daughter calls him the Beach Ball, displays most of the orange cat traits. He is the spoiled only cat of my best friend. Vinnie is insanely curious and likes to open up doors. Yet, despite his curiosity, he has no desire to leave his little urban balcony. He can sit for hours there and watch birds. Indoors, he likes to bat around socks and dustballs. He is so playful, even in maturity, that he can't be trusted alone with a Christmas tree. When Vinnie was a kitten, he was hell on wheels. My friend returned from a funeral one day to find that he had unrolled all the toilet paper in both her bathrooms. The first year she had Vinnie, he attempted to rearrange her Christmas ornaments in the middle of the night, causing the tree to come crashing down.

Vinnie has always been a biter. I'll be talking to my friend on the phone when she'll suddenly say"Ouch!" and I know that Vinnie has decided to bite the hand that feeds him. He also nips her ankles after she's been away on a trip, just to show her that he disapproves. He loves to sit in the hallway closet, and will do anything to get into it, knocking off coats that are on the door handle, and then fishing underneath with his paw to pull the door open.

When Vinnie arrived at my friend's house, he was a skinny stray. Only a year later, he weighed twenty-two pounds. He's the Alfred Hitchcock or Orson Welles of orange cats. Poor Vinnie—he hates to be the center of attention, and he always is, the second he walks into a room. His head is, of course, fairly normal, and so it often takes a few seconds for people to fully grasp his immensity. Then they gasp or cry out. Some dissolve into laughter just at the sight of Vinnie's corpulence.

It's a terrible fate for a shy cat. At one party with a Moroccan theme, a guest brought a tiny feline Fez and tried to put it on Vinnie's head. The little red hat with its ridiculous tassel stayed on for only seconds, but Vinnie's resemblance to Sydney Greenstreet, the obese actor in the movie *Casablanca*, was noted by all the guests. I felt very sorry for my pale orange buddy—he hates to lose his dignity and moped around my friend's place for weeks while he tried to recover from the Fez Incident.

Jasper the Ferocious

The same friend who ended up with Vinnie the Beach Ball also, at another time, owned Jasper. It's as if her life has been plagued by neurotic, orange cats.

Jasper was a piece of work. First, he managed to fall out of a third-story apartment window as a kitten, surviving dramatically and racking up veterinary bills in the hundreds. After that Jasper was never the same—something wasn't right in his brain, and the orange cat's natural tendency to bite a bit ran amok in him. He would charge at people, especially anyone who came to clean the house. My friend had an operation and needed home health care—because of Jasper's insanely violent attacks, two women quit within a day of each other.

Another of Jasper's charming traits was his ability to open cabinets and push dishes and foodstuffs onto the floor. He did it whenever it rained or it was sunny or he was bored or the moon was full, or, well, basically whenever he felt like it! Jasper was known for his mood swings, and at one point my friend was told that Jasper was severely depressed because of her own inability to control her anxiety.

Despite his lack of the usual orange cat friendliness, you had to admire Jasper's flamboyance. He lived to be almost twenty-five years old, on dialysis and skinny, but still snarling until the end.

Great Names for Orange Cats

O.J.

Orange

Mandarin

Navel

Judas

Tropicana

Red

Conan

Carrot Top

Bozo

Clem

Top Ten Ways to End up with a Cat

1. Open your door on a rainy night.
2. Grow catnip in your backyard.
3. Put up a bird feeder.
4. Develop an allergy to cats.
5. Buy new furniture.
6. Decide to feed a stray for "just one day."
7. Decide to let a stray stay inside "just for one night because it is so cold."
8. Buy a farm.
9. Get an apartment in the city.
10. Look into that box marked "free kittens."

8

Black Cats

> If a black cat chooses to make its home with you, you will have good luck.
>
> —English superstition

Black cats at-a-glance

Size: Small to medium

Tendency toward obesity (on a scale of 1 - 10): 2

Sex: Both flavors

Color combos: Black, with occasional tiny white spots, sometimes barely imperceptible white or chocolate tips

Peculiar traits: Usually very vocal, very athletic and "antsy," have a tendency to wander more than some other colors of cats

Craziness (on a scale of 1 - 10): 9

Famous black cats: Piewacket in *Bell, Book and Candle*, Salem on *Sabrina the Teenage Witch*, Edgar Allan Poe's "The Black Cat," the black cat on the Schwarze Katze wine label

Black Cats: Back to Basics

Verbal, slinky, nutty—black cats have a certain *je ne se quois* that has not gone unnoticed throughout history. For some reason they have been singled out as particularly evil, and yet anyone who has ever owned a black cat knows that the opposite is usually true—they are sweethearts.

Perhaps the witches of old chose black cats for their companionship qualities and not at all because they were scary or messengers of Satan. The neuroses of black cats can make them difficult, but, then they have a lot to be neurotic about! For centuries, terrible people have persecuted black cats, killing and torturing them in large numbers. And, more recently, black cats have lost out in the Darwinian alley cat struggle against the automobile. I noticed, growing up, that our black and gray cats were much more likely to be smooshed on the highway than their lighter-furred friends.

Black Cats I Have Known and Loved

Willy, the Cat-Whale

Willy got the name from the movie about the whale, *Free Willy*. My daughter, Kelly, thought that the piteous cries Willy uttered were similar to the Orca whale Willy's communications with his family in the ocean. He also arched his back as he wailed, which reminded her of the whales jumping out of the ocean (keep in mind that she was only five years old at the time). I always laughed when I thought of Willy as a whale, because he was one of the tiniest cats I've ever owned. He was a classic black cat who might have posed for all those Halloween silhouettes. He had a very skinny profile and the world's smallest head. Yet he was packed with personality.

Willy must have had some Siamese in him—many black cats do—because his voice was remarkable. And he seemed into unrewarded bouts of self-expression—he was altered, so it wasn't as if he was trying to attract a mate or fight other males. He just liked to sing. He would get on a roll and big, powerful, chanting syllables would come out of his mouth. It seemed as if he dragged them out of the diaphragm and vocal chords of a hundred-pound cat.

He had spirit, too. Willy was the only one of our cats who would ever wrestle with our Jack Russell terrier. And when he did, he made dramatic noises to try to distract the dog. Yowling and spitting, he would pounce, always going for a surprise attack. Even as a puppy the dog weighed a good few pounds more than Willy, so he always pinned him down. But Willy would continue to make so much noise that the dog would eventually let him up in disgust. (If they had been cartoon animals, the dog's paws would have flown up to his ears!)

Cat, the Wanderer

Cat was our neighbor's pet. She liked to hiss and spit at dogs and humans, so she was often seen in that classic "Halloween Black Cat" pose—arched back, puffed fur, curled tail. She was remarkably youthful, staying all black until she was nearly twenty years old. She was also, like Willy, very athletic. I would look outside my upstairs window and see her scaling walls and climbing fences with total abandon.

Cat had that traditional black cat wanderlust thing going on. She just couldn't sit still. And, like Willy, she liked to scream and moan. Everyone in the neighborhood knew her. There are two parks near our house, both a few blocks away, and Cat would show up at each of them frequently. It's a wonder no one ever snatched her or called the animal shelter.

Cat had a game—she would loiter around my front door, wait for me to step out, and scurry inside our house. I'd run in after her, trying to catch her, and by the time I found her she would already be in a bedroom preening. She always looked very surprised to be evicted.

Monty, the Dim-witted Boomerang

I already mentioned my theory about long hair and cats: the longer the hair, the dumber the animal. Monty was a stellar example of this rule. But, wow, was he pretty! Technically, he wasn't exactly a black cat, although in most light he looked black. He was a product of a slight dilution, and was essentially a chocolate-colored longhaired cat. People would ask me all the time if he was a Persian or a Maine Coon cat, but he was just actually a lucky genetic accident, an alley cat that happened to look classy.

Monty was dumped in our backyard with his sister Edie and two other kittens (see the calico chapter). Although he was the male of the brood, he always seemed very feminine, and was never very good at hunting or fighting. He had a nice walk, though—a kind of regal strut. And his grooming was legendary. He seemed to spend about nine or ten hours a day licking and arranging his magnificent fur. Early in his childhood he discovered that he enjoyed sitting in hats, particularly straw hats. I have many pictures of him in his hat-nests, and I particularly like the one of him in his favorite sombrero.

Monty was a bit dog-like—he fetched objects and begged for food. He was particularly fond of shrimp shells.

When I was pregnant with our daughter, Monty disappeared. Everyone said that cats could sense changes, and that he had left in anticipation of the birth. "Bull!" I said. Monty wasn't even smart enough to know that when you put suitcases out, you were going on a trip, so I couldn't imagine he would catch on to my pregnancy. I put up signs all around the neighborhood, and in one eerie episode actually got a call from someone who had found a cat they presumed to be Monty. I walked out onto a fire escape and greeted a beautiful chocolate-colored longhaired cat, but it wasn't Monty (this made me wonder, though, if within a half-mile radius, there weren't lots of similar siblings sired by the same Tomcat). Time went on, and I was always very sad that Monty didn't return. One day, almost two years later, I was squatting near the ground, weeding the garden near the fence in our little backyard, when I heard the plop of a cat landing from atop the railing. I didn't even look up, since cats flying through the air and landing on valuable plants are a regular event in my life. But when I did turn around to throw away my weeds, I was face to face with a cat that looked as though it was from Mars. The poor thing was bald except for a tiny tuft of chocolate brown hair at the very end of its tail.

"Monty!" I screamed. My poor little guy was ugly almost beyond recognition, but he had found his way home. A visit to the vet revealed that he was so flea infested that he had scratched or bitten off all his fur. And he no longer possessed his famous nails—whoever had stolen him had declawed the little critter.

It took months, but Monty regained his long, glossy coat, and he continued to dwell in our little yard, even climbing fences without his front claws. Yet I always wondered where he had been. I imagined that whoever had found or stolen him felt lucky to get such a toney cat for free. Then he probably got fleas, and as soon as he became really ugly, they threw him out. I hate humans who do that!

Shlomo, the Bad Boy of Spruce Street

Shlomo was the worst cat I've ever owned, and one of the most charming. When he arrived in my backyard he was an unaltered stray. My hallway and basement still carry his scent to this day, and he's been gone for over seven years.

I recoiled in horror when I first saw Shlomo traipsing through my backyard. He was a classic black Tomcat, with multiple chunks bitten out of his ears. His tail was partially severed, hanging from a thread. Several days later the tail was gone, with a bloody stump left behind. We dubbed him Stubby, and kept hoping he would go away. He was very mean to my cats, beating them up regularly; in general, he terrorized any feline who deigned to come within a foot of the yard, even if they did belong there. I've never before called the ASPCA to come get a stray cat—I've even set up my backyard with a special "cathouse," a wooden doghouse I bought at a Mennonite flea market to shelter nomadic kitties. Yet, I had just started to harden my heart for this impossible case—a nasty cat with no redeeming qualities and an appearance that made me slightly nauseous. On the very day that I decided I had better call some shelter folks and get them to take Shlomo (Stubby at the time) away, I was dawdling in the backyard, reading the paper on a beach chair in the sun. Without warning, this hideous, beaten-up, tail-less alley cat jumped into my lap. Shlomo looked into my eyes and began purring, pushing his nibbled ears toward my neck. I knew I could never send him to certain death at a shelter. (Who besides me would be stupid enough to adopt him?)

So, Stubby went for his sexual alterations and then returned, pretty much as mean as ever. But his tail healed, and his coat became glossy and gorgeous. His chewed ears also smoothed out a bit. My husband thought that the name Stubby was politically incorrect, as it called attention to the poor guy's tail-challenged state. So, since we had decided that we would start naming all our male cats with "O" names—Leo, Hugo, and Cosmo—this new, improved black cat became, permanently, Shlomo.

Shlomo was big and muscular and totally devoted to me, and to our daughter, Kelly. She was about two years old when Shlomo first came to live with us, and at the preschool she attended down the block, they thought she was talking about me all the time. "Mo-Mo!" she would say, several times a day, to no one in particular. Her teachers assumed it was her special word for Mommy, but she was really trying to engage them in a conversation about her favorite cat.

Shlomo never really did get along with any of my other male cats. He particularly picked on Monty. I'd hear the hissing and yowling in another room, and then go in to find a substantial hunk of Monty's fur lying on the floor. Eventually I had to confine Shlomo to the basement at night so that Monty wouldn't be completely scarred and bald. It became our nighttime routine. I would catch Shlomo, and he would obligingly go limp in my arms, resigned to his fate of being thrown down the basement steps and into solitary confinement. In the morning he was always waiting patiently at the top of the stairs to resume his role as Bully of the House.

Shlomo was an amazing athlete. He could jump so far that I always wondered if he could jump out from our roof to our yard. He was always roaming the roofs of the neighborhood.

I became very fond of Shlomo, fonder than I even realized, since he was only on this mortal coil but a year and a half before we lost him. He died as he had lived—dramatically. I walked into our yard early one evening and he was sprawled across the pavement. I put down my hand to pet him under his chin, and he turned toward me and opened his mouth. Instead of his usual meow greeting, he gasped and then was suddenly still. I was in shock. I picked him up and he was limp. "No, no, no," I cried, carrying him into our house. I put him on our wooden floor, and, one by one, our other six cats walked past and sniffed him, saying their final, and rather joyful, good-byes to Bully Boy.

As the night progressed, I was not in my right mind. I cried. I sobbed. "He was like every bad boyfriend I never had!" I said to my perplexed husband. I realized that I'd adored Shlomo because he was so unrepentantly bad—like those juvenile delinquents you see at amusement parks with cigarette packs rolled up in the sleeves of their t-shirts. I had loved the little smirk he would give me as I exiled him to the basement every night. It was almost as if he was saying, "Yeah, I'm bad, but you know I'm what you need, lady!"

Then things got worse. My agitated husband decided he would bury Shlomo that very night in our tiny, urban backyard—maybe all my boyfriend talk was making him jealous. He dug a hole and wrapped the magnificent corpse in some old towels. At this point I was flinging myself about the room in a very Irish Banshee kind of way, wailing "Mo-Mo! Mo-Mo!," just like our two-year-old. Eventually I fell asleep, but awakened extra-early, as did my mate. We looked at each other and said the same thing, "What were we thinking? That yard is too small for his body!" So my husband went out and unburied Shlomo. I refused to even look. He took the cat's corpse to the vet, where he was cremated. But as my husband was getting into the car, I kept yelling at him, "Tell them why he's dirty! Tell them we loved him, and we didn't mean to make him dirty!" The kind vet told me that even young vigorous cats such as Shlomo have heart attacks, and from my description, she thought that he had suffered a massive coronary.

Alas, poor Shlomo, he really *was* the bad boy of Spruce Street.

Great Names for Black Cats

Blackie

Midnight

Tar Baby

Coal

Truffle

Noir

Licorice

Caviar

Names Based on Physical Attributes

(Be careful: Your cat can get older, and appearances can change!)

Tiny	Shorty
Midgie	Crybaby
Gimpy	Snoopy
Fuzzy	Boots
Stubby	Zorro

Human Names that are Good for Alley Cats

Lucille	Felix	Duncan	Cassius
Mildred	Esther	Jebediah	Tom
Bernice	Fred	Treat	Daisy
Otis	Klaus	Hugo	Liza
Homer	Ophelia	Violet	Basil

Calicos and Tortoiseshells

"A calico or a tortoiseshell is like a Hershey bar with almonds—half sweet and half nuts."

—Betty Crane

Calicos and Tortoiseshells at-a-glance

Size: Small
Tendency toward obesity (on a scale of 1 - 10): 3
Sex: Almost always female
Color combos: White, tan, brown, orange, black, (the tortoiseshell is just a version of the calico without large white spots)
Peculiar traits: Shy, unpredictable, extremely neurotic, or, as many calico owners say, "just plain crazy"
Craziness (on a scale of 1 - 10): 9
Famous calico and tortoiseshell cats: The Calico Cat of nursery rhyme fame

Calico and Tortoiseshell Cats

The absolute nuttiest breed in the color kitty zodiac, these almost-always-female specimens have quirky, introspective personalities, and are often one-person animals. In my survey of vet's offices, I would frequently hear, "Well, I can't be sure of all the different personalities, but I can tell you one thing—calicos are crazy!"

What is it about these cute, patchwork felines that makes them nuts? I have an amateur theory that the orange color gene gives cats a certain neurotic craziness. It surfaces in the male, orange cat, and in the female it manifests itself in the calico, or tortoiseshell.

Calicos appeal to people who truly love a cat that is all cat, one that doesn't particularly care to please humans that much. You won't find a calico socializing with huge groups of humans, and especially not children. Even though strangers will be attracted to a calico's beauty, she will run away from the attention.

It's hard to generalize about intelligence in cats, but I have known a number of rather dim-witted, addled calico girls. Lucille was one of them. Her "mom," Anne, used to put her outside on her apartment balcony on fine spring days, and then close the sliding doors for a while. Lucille would stand and look back and forth at the door, and Anne would then open it. But Lucille would fail to grasp that the glass was no longer there. Anne would have to lift her through the imaginary glass each time. "It would be one thing if this was only once or twice," says Anne, who claims that Lucille was absent the day God gave out brains to kitties, "but it happened every single time, for years!"

Calicos can be fierce, especially if crossed. My vet calls them schizy, adding that, "Their owners always say they are so nice, but when I examine them they fight me like crazy and often have to be restrained." Others report this schizophrenic behavior. A friend of mine, Betsey, was asked to take care of her sister-in-law's calico, Hannah. From the moment Hannah entered the house, she began snarling and clawing at furniture and humans. Betsey had to put on her old field hockey gear and huge asbestos gloves to handle Hannah and barely managed to get her upstairs to the attic, where she remained for months. A few years after Hannah had been returned, Betsey attended a dinner party at her sister-in-law's place, and Hannah remembered her. "I couldn't believe it," she told me. "The damned cat ran over and stuck its claws into my rear end, up through the chair seat! And meanwhile everyone there kept saying what a sweet cat she was."

Calicos and Tortoiseshells I Have Known and Loved

Kitty, the Wheezing Princess

Kitty started out her life as Princess Calico Kitten. And, indeed, she was a princess—aloof, with a regal walk that could shake the confidence of any feline interloper. She was the only alley cat I've owned who actually looked good in a pink collar with little bells on it. She remained impossibly skinny throughout her life, like a kitten movie star, and looked very chic in her faded coat of many colors. (Calicos really run the gamut from bold oranges, blacks, and browns to subtle tans, grays, and amber.)

Kitty ruled the roost as an only cat for four years before our household became a multiple cat dwelling, so I always thought that was why she never warmed up to her other furry sisters and brothers. After Kelly (see below) arrived, Kitty retired to the cellar of our two-hundred-year-old farmhouse for a period of reassessment that lasted nearly a year. She spent most of her time in a tiny crawl space way in the back. Anyone who disturbed her was cut down to the quick by her disdainful expression. Kitty was really my sister Debbie's cat, and only Debbie could really relate to her fully. They were both shy yet fully confident that they were better than anyone else. Even during her period of cellar exile, Kitty would sometimes deign to sneak up in the dead of the night to sleep in Debbie's room.

My mother and sister had rescued Kitty from a terrible pet shop. Kitty was already four months old at the time, and well past the fuzzy kitten stage. Whether it was because of the unsanitary conditions in the shop or a hereditary condition, Kitty suffered from terrible asthma, not in keeping with her dainty appearance. She was, most times, a raucous mouth-breather. Often at night in our old, spooky house, my little sister and I would be convinced that a monster or burglar was coming up the stairs or down the hallway, only to find out that it was Kitty wheezing her way around the house. I felt sorry for her—she always seemed mortified to be asthmatic. Over the years we tried all sorts of medicines and treatments, but nothing worked.

Sometimes I worry that calico cats are less sturdy than most alley cat types. Kitty's other health calamity occurred when she was about five—a hit and run accident in which the automobile got the best of her. Kitty's leg was shattered, and had to be put back together with pins. Worst of all, for a vain girl such as herself, most of her lower body was shaved and stitched. She spent eight weeks in a cage in our dining room because she wasn't allowed to move around. Day after day she would sit there as we ate our meals, glaring at us through yellow half-slits. Whenever one of us would try to give her affection, she would turn her back and stick her tail out of the bars.

Oddly enough, Kitty adored dogs. She would often sit with our Cairn terrier, Bonnie, and purr and wheeze. The two animal friends always reminded me of the Calico Cat and Gingham Dog of the famous nursery rhyme.

Edie

Edie was a dumb girl. You would look into her eyes, and there was nothing there. And yet, she was beautiful—a classic medium-haired calico. She was nearly a true tortoiseshell, but she had white feet, and a splendid white chest. She arrived as a stray kitten with her brother Monty, and was pretty much subordinate to him. The two were inseparable and great grooming partners.

Edie was a water cat. She loved to sit on the sink and play with the wet stuff coming out of the faucet. She would wait until after you stepped out of the shower and jump in to chase the leftover drops around. In the backyard, she would act like Narcissus, staring at her own image in a puddle and tapping the top of the water with her paw.

Like many kitties, Edie had one great adventure in her life, and it wasn't a pleasant one. She tried to kiss an automobile early one April evening. My husband and I were out at the time, and fortunately a kind neighbor had brought Edie to the vet hospital, where they were able to keep her alive. She stayed there for weeks, racking up the bills, until they called me and said there was nothing else they could do for her. She simply would not eat. I could choose to have her put to sleep, or I could take her home and see if she would start eating on her own. I chose the latter approach.

Well, as I said, Edie was dumb before a car ran into her head, but afterwards she was even dimmer. I spent hours with her, pushing her nose and mouth into tuna fish. She never seemed to make the connection between my actions and the food in the dish. She didn't much like having some smooshed stuff on her face, though, and would try to hobble away whenever she saw me coming with the dish. After a week I was about to give up and send her off to kitty heaven, when she suddenly got it—food! I'm hungry! Food in the dish! She took a bite and snuffled, then took another. She lived for years afterwards and never went near the street again.

Kelly, Woman Warrior

It was a dramatic scene—our dog Bonnie was growling and snarling and lunging in front of the stone wall near our barn. My dad came running, fearing that she was once again tangling with a woodchuck, but when he got there all he saw was a tiny calico kitten hissing and spitting at our little terrier. All of her littermates were dead, having been grabbed by the neck and shaken by our little dog (cairn terriers are trained to kill rodents in big rock piles, so it was hard to blame the dog). My dad quickly scooped up the fuzzy welterweight and brought her into the house, where we put her on the kitchen table and did an amateur examination. She seemed totally intact.

Even years later, my dad would talk about this little cat's courage. "She was determined to fight until the end," he said.

My mother wanted to call the kitten Kelly, after her maiden name, because it was a girl's name she'd always wanted to use. (All of our animals were females, and they all had names of daughters she would have had—Amy, Bonnie, etc.). And she thought that Kelly was a feisty Irish name, appropriate for the last survivor of a battle defending her homestead. Kelly means "woman warrior" in Celtic.

We never found out where Kelly's mother had gone, and, as far as I know, she never reappeared. Perhaps the tragedy of losing all her kittens was too much for her to bear. Interestingly, as soon as the little cat entered our house, the dog never bothered her again. This shows how specific breeding can be—if there was a little funny critter in a pile of rocks, Bonnie would kill it. If, on the other hand, the critter was sitting on our couch, it was okay.

Kelly was a pretty critter, that's for sure. She was a dark calico—bright orange mottled with a chocolate brown and black and a little white. She had a broad head and a nice, compact body. Her whiskers were a magnificent feature and the source of a great mystery in our family. Once, after a New Year's Eve party at our house, we discovered a clumsy Kelly walking oddly and knocking into walls. We feared the worst: brain tumors, epilepsy. But closer inspection revealed that someone had cut off her whiskers, right down to the nub. My godparent's three rambunctious boys were prime suspects, but we never discovered who did it. Kelly's whiskers gradually grew back, and she regained her natural agility.

Kelly never meowed. I never realized how much this perplexed my mother until one day I came into the house and noticed that my mom and my sister were holding down the lid on a cardboard box. "We put Kelly in this box, and we want to see if she'll meow to get out," said my sister, who was twelve at the time. My mother, looking a bit embarrassed, said, "Well, it's sort of like a scientific experiment. She's been in there five minutes already, and not a peep!"

Great Names for Calicos and Torties

Bridget
Raku
Tapestry (Tappy)
Delilah
Sarah Bernhardt
Medea
Zelda
Mata Hari
Lizzie
Madonna
Cher
Gingham
Betsy Ross
Patchwork
Turtle
Rapunzel
Taffy
Ratatouille

Names for Multiple Cat Households

Often people will adopt alley cats in pairs—here are some good names for brothers and brothers, brothers and sisters, or sisters and sisters.

Boris & Natasha
Ike & Tina
Calvin & Hobbs
Mutt & Jeff
Frankie & Johnny
Orville & Wilbur
Naomi & Winona
McCabe & Mrs. Miller
Fred & Ethel
Samson & Delilah
Lancelot & Guinevere
Catherine & Heathcliff
Melvin & Howard
Cyrano & Roxanne
Procter & Gamble

Simon & Schuster
Sears & Roebuck
Yogi & Boo Boo
Flotsam & Jetsam
Nichols & May
Popeye & Olive Oyl
Sacco & Vanzetti
David & Bathsheba
Siegfried & Roy
Penn & Teller
Don Quixote & Sancho Panza
Leopold & Molly (Bloom)
Jake & the Fatman
Peter & the Wolf
Homer & Marge (Simpson)

Names for Multiple Cat Households (continued)

Donald & Ivana
Gertrude & Alice
Sonny & Cher
The Captain & Tenille
Bob & Ray
Chad & Jeremy
Martin & Lewis
Rowan & Martin
Porgie & Bess
Gomez & Morticia
Lucy & Ricky
Gable & Lombard
Ozzie & Harriet
Darren & Samantha
Adam & Eve
Rhett & Scarlett
Lady & Tramp
Cain & Abel

Scott & Zelda
Lyndon & Lady Bird
Jack & Jackie
Donnie & Marie
George & Gracie
Beauty & Beast
Barney & Betty
Lewis & Clark
William & Mary
Dow & Jones
Dun & Bradstreet
Flora & Fauna
Surf & Turf
Chip & Dale
Amos & Andy
Bonnie & Clyde
Dick & Jane
Sally & Spot

Names for More Than Two cats

Tom, Dick, Harry

Kukla, Fran, Ollie

Bob, Carol, Ted, Alice

Mo, Larry, Curly (optional add-on: Shemp)

Groucho, Harpo, Chico

Wynken, Blynken, Nod

Flopsy, Mopsy, Cottontail, Peter

Gray Cats

At dusk, all cats are gray.
—popular English folk saying

Gray Cats at-a-glance

Size: Small to medium
Tendency toward obesity (on a scale of 1 - 10): 9
Sex: Both flavors
Color combos: Light gray, dark gray, gray with white boots, chin, or crotch area
Peculiar traits: Collect shiny things, like to sit on radiators or in the sun, somewhat shy
Craziness (on a scale of 1 - 10): 6

Gray Panthers: The Easygoing Grays

I have a plan that I will be pitching to the Department of Energy: solar cat power! Just put a bunch of gray cats on sunny windowsills and let them soak up the rays. Then make them walk around, dispensing heat.

Grays are mellow cats, and tend toward obesity. It's as if, with the dilute gene making them gray rather than black, it also reduces their neurotic nature. A gray cat would make a very good companion for an older person or a shut-in—they are content to keep close to home and hearth.

Gray Cats I Have Known and Loved

Sonny, the Fat Lady of the Opera

My wonderfully obese gray cat Sonny was a magpie, like many grays. She would steal watches, earrings, and other shiny objects and stash them in a hoard behind the refrigerator. A friend of mine who stayed over one night was dismayed to discover that Sonny had dragged her brand new watch, a gift from her father, down three flights of stairs. Fortunately, the watch still worked when we found it on the couch. I often wondered if I could take Sonny with me to gather jewelry at parties just as some thieves take monkeys into a crowd to pick pockets. Unfortunately, though, Sonny could only snatch things that were already off of their owners. As talented as she was, she couldn't manage clasps. All cats enjoy Christmas ornaments, but Sonny was a pro at taking them off the branches and scattering them throughout the house.

Stealing jewelry was only one of Sonny's many eccentricities. For years she had a wild friend in our urban backyard—a possum. We called him her boyfriend. It was quite a sight to look outside and see those two gray creatures sitting side by side, silently communing. Perhaps gray cats like the companionship of other species—I know Sonny did, and so did several other gray cats I've known, although they were usually more conventional in their choices and went with dog friends.

I don't know why gray cats are so prone to corpulence, but it is a well-known fact among gray cat aficionados. I've seen fat gray cats in nearly every European country, and all over the United States.

When Sonny was a young cat, I took her to a Korean veterinarian who spoke very little English. I tried to ask him about my cat's weight, and how I could keep her from getting too heavy. The language barrier made it impossible to discuss. The vet just looked at me and said. "Oh. Cat not fat!"

A year later, I took Sonny for a check-up and he said "Ohhhh. Cat fat!"

By then it was too late.

Eventually Sonny became fat enough to cause friends to gasp as she entered a room. She was a small-boned cat who really should have weighed seven pounds or less, but she tipped the scales at over twenty. At one point I dressed her in a fake pearl collar so that she looked like a fat lady going out to the opera.

I tried for years to get Sonny to lose weight. If I starved her, she would just stop moving. One time over the holidays, we had a disaster—our pipes and radiators froze and burst. We had given the friends who were supposed to look after Sonny the wrong key to our house, so after her dry food ran out, Sonny was without for several days. But she was resourceful. She found some Christmas cookies up on a counter and gnawed at them like a squirrel until they were all gone. In fact, the idea that there was no food in her dish never stopped Sonny from seeking sustenance. She was an excellent hunter, moving swiftly despite her stoutness. She killed and ate roaches, mice, voles, and, in one instance, a hapless bat. Anything to supplement the diet!

If wild game wasn't available, Sonny would also attack and dismantle garbage containers. One night my husband and I heard terrible shuffling noises from the kitchen. It sounded as if all our possessions were being rearranged by burglars downstairs. Trembling, we called the police from our bedroom. They arrived only minutes later. No intruders were in the house, but Sonny was seen in the kitchen sitting contentedly licking her paws next to a huge, opened bag of garbage she had dragged from the pail.

Clyde, the Dog Lover

Clyde was a female, gray kitten who loved dogs. She got her name from the popular movie of the 1970s, *Bonnie and Clyde*—we just happened to have a cairn named Bonnie at the time. Yet it was Amy, our beloved St. Bernard, who became Clyde's constant companion. I have many pictures of them together, with the longhaired Clyde curled up in Amy's tail or eating from Amy's dish while the huge dog waited her turn. Their relationship reminded me of those touching stories about huge Gorillas in captivity at the zoo who adopt kittens.

Clyde never really came into our house—she was a true-blue feral cat. One advantage of staying outdoors was the spectacular winter coat she developed. Every fall her hair would go gangbusters until she had long, glossy, gray fur that dwarfed her skinny body. I remember how proud I'd be when people asked if she was a Maine coon cat. It's a good thing she never did come in the house, though, because in the spring the opposite happened—- all you had to do was touch Clyde and you would be left with a mighty handful of gray fur. She often looked quite moth-eaten as her shorter summer coat came in. She enjoyed being petted, but she also enjoyed her freedom, and was a bit skittish (she had been thrown out of a speeding car as a baby).

Great Names for Gray Cats

Smoke

Panther

Ash

Flannel

Phil (for Phil Donahue)

Bill (for Bill Clinton)

America (for the Graying of America)

Grayson

Grace

Sophie

Kate

The Brain

Interview with a Cat Lady

Betty Crane calls herself a rescuer, but also doesn't mind being known as The Cat Lady. In the past fourteen years she has rescued hundreds of kittens and grown strays from the Philadelphia streets, and placed them in good homes. She works alone and with friends, and is not part of any formal organization. When I asked my vet who would know the most about alley cats of different colors, she didn't hesitate. "Call Betty Crane," she said. "She has seen more cats than almost anyone I know!"

Q: How did you get started rescuing stray cats?

A: In 1987 I was walking my dog past a cemetery and noticed a litter of tiny kittens. They were so cute, I just had to help them. This was quite an unexpected thing for me: my husband had a cat, but I'd never really warmed to it. It was his cat. And then I had my dog. But that litter of kittens changed everything.

Q: Did you keep any of them?

A: Yes, the male longhaired tabby. I placed all the others, even the one we called Jesse because she was like Jesse James—she was impossible to catch. Now I realize that she was just your typical stubborn calico.

Q: So you do notice differences in personalities, according to color?

A: Oh, yes. For example, the orange gene, when it appears in females, particularly the calicos and the tortoiseshells, produces a very strong-willed cat. Calicos can be very sweet, but then they can be very nasty, too. They also have a nutty streak. And they're not that good with other cats. They're better off in a one-cat household.

Q: But what about orange males?

A: They are true sweeties. Male, orange tabbies are friendly, and outgoing. And yet they are also more dominant than other-color, male cats. They spray more frequently. I've noticed that you can have only one male, orange cat. He'll always be the alpha cat. I recommend that if people adopt one, then they adopt only female cats after that. And here's something interesting: if the orange color is light in a male, then he will be mellower. That's true of the female orange-colored calicos, too—the ones we call ash, the dilute-colored ones, are easier to live with.

Q: You mentioned, though, that orange cats have special health problems.

A: Yes. They seem to get diseases of the urinary tract more frequently. And they get those black spots on their gums.

Q: They can be biters, too?

A: Yes! I hear the funniest stories. One woman told me she had to wear winter boots all the time because her orange cat was biting her. My take on that is, you can train cats—you don't have to put up with that kind of stuff!

Q: What are your other experiences with different color alley cats?

A: Well, I've found that totally black cats are great—and my theory is that they usually have some Siamese in them, so they are friendly and outgoing. But when you have a predominantly white cat with black spots, watch out! A lot of times they're insane. I've had a lot of black-and-white cats attack people. Gray cats, on the other hand, are peaceful and loving. A lot of times it's hard to know that a person even has a gray cat—they don't come around for visitors. But when they are young, they can be very energetic. We had one called Nut Case who loved chasing light. All you had to do was shine a flashlight on a wall and he would chase it forever.

Q: Are all personality pluses or minuses color-related in cats?

A: Oh no. What most people don't understand is that about 80 percent of cats are shy and not that great around people. They can't be the type of pets people want. It's all genetic—there must be a shy gene, too. And a lot of people will just assume that a shy or standoffish cat was abused as a kitten. But I raise many kittens—I don't let them go to homes until they are twelve weeks old—and I see a lot of shyness and aloofness. It's a natural cat trait. Yet cats can be trained. I train my kittens to jump automatically into a cat carrier. It's easy, if you start early enough. You would be impressed—I can just say, "Go into your house!" and my kittens hop right into the carrier. None of that fuss. Let's face it, every cat is going to have to ride in a carrier at some time or other. It's a good skill to teach them.

Q: What is the most surprising thing you've learned about alley cats?

A: I've found from doing rescues in urban areas that there is a myth about the tabby cat (the gray and black striped cats). People think that they are the natural alley cats who belong outside. I've heard this lots of times, as if tabbies are the standard, boring cat not even worth owning while the other colors are special. I don't understand it. Tabbies are friendly and outgoing, and I think they're beautiful. Why, I just rescued a boy tabby kitten that has two pure white sisters, and I think he is more beautiful—you should see his pattern. But people don't see the tabbies as special. Just a few weeks ago I was scheduled to pick up a litter of five kittens, and the person who had found them said, "Well, you'll only have to take three of them." When I asked why, she said, "Well, two of them are alley cats, the tabbies, so I can just put them out on the street."

Q: How many cats do you own?

A: Eight.

Q: Any favorites?

A: Well, I don't play favorites that much. They're all different. But I do love my Taffy, the tortoiseshell. Torties love to give kisses. I can say her name, and she runs up my leg, puts her front legs around my neck, and buries her face in my ear!

White Cats

To see a white cat on the road is lucky.
—American folk saying

White Cats at-a-glance

Size: Small to medium
Tendency toward obesity (on a scale of 1 - 10): 4
Sex: Both flavors
Color combos: White!
Peculiar traits: Verbal, witty cats, most definitely lapcat material, voted most likely to unwind the toilet paper
Craziness (on a scale of 1 - 10): 9
Famous white cats: The villainous white cats in the James Bond and Austin Powers movies

White Cats: Flashy and Fun

Picture this: An Irish monk in the year 800 A.D. sits in the library of his monastery, doodling in the margins of a long manuscript he is copying. He is writing, in Gaelic, a poem about the glorious personality of his dear friend, Pangur Ban. Who is this "person" he is praising? His white cat!

People are fanatical about their whites, and with good reason. White cats have a human-like personality. They're the slightly eccentric relative whose jokes you always enjoy. They are high-spirited, and fun to be around. Every white cat I've known craves close physical contact—some won't even let you read a newspaper without jumping onto your lap.

White cats are aesthetically pleasing. The paleness of their fur cuts through the landscape and makes them like a daily dose of moonlight. They are, I think, the most exotic of alley cats, and some of the most pleasing around.

Whites I Have Known and Loved

Boo Boo: the Chimney Sweep

When I was little, as I was falling asleep, I'd often be awakened by Boo Boo, our crazy, white cat. She was small and feisty, with a round face. She wasn't deaf, but she was totally nuts. Boo Boo never met a human lap she didn't like. She also was addicted to board games—we could never play Monopoly or Parcheesi or Clue without Boo Boo first perching herself at the side of the board and then, eventually, in the middle. We would just lift her tail as we moved our playing pieces.

At night, Boo Boo prowled the slate rooftop of our old farmhouse, crying at the windows and generally making a nuisance of herself. I always wondered how she could keep from falling off the slippery slate shingles. One night her wanderings went too far—she fell into the chimney. There must have been little footholds on the way down, because the next morning, in the family room, we heard her wailing from somewhere within the big stone fireplace.

"Where the heck is that cat?" said my father, sticking his head into the fireplace opening and as far up the chimney as he could. "Boo Boo!" we all yelled. Almost on cue, as soon as he withdrew his head, Boo Boo came flying down the sooty orifice, landing on all fours between the fireplace andirons. "Meow," she announced, then composed herself and walked nonchalantly into the room, holding her tail high in the air. We all burst out laughing, because Boo Boo was now a gray cat, her magnificent white fur coated with chimney soot.

Caspar: the Weird Ghost

My husband was very sick in the hospital, and another cat was the last thing I needed (especially not a crazy white cat that screamed and whined). He was young, and, I figured, probably unaltered. Every summer we end up with lots of young male cats that were once cute kittens. Then the college students go home to their parents, and suddenly the kittens aren't as cute anymore, and they stay here. Not only that, but they stay here and create other cats, since by the time the students dump them they're usually on the verge of sexual maturity.

This one had come in through the upstairs window we keep open for our cats. I had heard him screaming out in our yard, but after visiting my husband all day in the hospital, I didn't have the strength to care about him. I got into bed and was sucking on a juice box. For one demented moment, I thought the squeaking noise was coming from my drink. But then I flung my arm over the side of the bed, and before I knew it, a small feline took a bite out of my finger.

Caspar was a pain in the butt from the moment I met him. And I only got to own him for a few weeks! He was a real half-pint, one of those white cats whose fur is the length of a mouse's hair. His belly was very pink. But even though he was tiny, he made an impact, verbally and physically. He seemed to be everywhere, demanding attention and food. And he kept up the yowling. My mother, who loves all cats, said, "That one is really annoying."

And yet he had something about him. At night he would curl up in my elbow, purring away. He was awfully cute, darn it.

Of course, Caspar was also a kitty teenager, dealing with his newfound sexuality, which included spraying his urine everywhere. So I knew he had to go to the vet (or, as we call her around our household, "the yucky lady") to be altered immediately. I figured it would cost me about seventy-five bucks, and then we would see what the next step was.

Gee, was I wrong. My vet called to say that in the middle of Caspar's... ahem...procedure she discovered that he had a hidden testicle. I don't exactly know what that means, but I know that it was complicated, and it cost me two hundred and fifty bucks. Then the noisy little guy came home. I kept him locked in our bedroom for two days, but I should have named him Houdini. We still don't know how he escaped and got out the window, but Caspar never darkened our doorstep again with his little white hide. "Whoa," he must have said to himself. "The food was good, but I don't much like them tinkering with my nether parts."

Since we never saw him again, I decided that he had returned to some household that now had the benefit of having a neutered cat. I got philosophical about it—my actions had cost money, sure, but at least it meant that there wouldn't be any Caspar Juniors yowling around.

Great Names for White Cats

Snowball

Elmer

Powder

Albion

Alba

Tennis

Blanka

Bianca

Bread

FAQs on Alley Cats

Why are there so many alley cats?

The average female can have three to four litters per year, averaging four or five kittens per litter. You do the math! (Assuming that each of her offspring stays unneutered, a cat mom can be responsible for twenty thousand felines in five years.)

Why are there so many alley cats wandering around the streets (and alleys)?

Only one in nine of all cats has an owner.

What is the largest litter of kittens recorded?

In 1970 a British alley cat gave birth to nineteen kittens, four of which were stillborn.

Why do alley cats in the same litter have so many colors and personalities?

One in four pregnant cats carries kittens fathered by more than one mate. During her fertile period she may mate with several Tomcats, each fertilizing different eggs each time. This explains why litters are often so varied in color, looks, and temperament. And, I suppose, why a house of ill repute used to be called a cathouse.

What was the one single invention that spurred the popularity of the cat as a house pet?

Kitty litter, which was invented in 1947 by Edward Lowe in Cassopolis, Michigan. Before this nifty clay pellet product, cats were often kept outdoors.

Why does my cat shed so much?

There are approximately sixty thousand hairs per square inch on the back of a cat, and about one hundred and twenty thousand per square inch on its underside. Apparently, it is the longer daylight hours as spring approaches and not the warmer temperatures that trigger molting.

How fast can cats run?

A cat can run at speeds of up to twenty-seven miles per hour. Although this is slightly faster than a human sprinter, it is quite a long way behind certain breeds of dogs such as the greyhound, which can reach forty-one miles per hour.

How long can a domestic cat live?

The average life span of males and females alike is from fifteen to seventeen years. The oldest cat on record is probably Puss, a tabby owned by a Mrs. Holway of Devon, England. He was thirty-six years old when he died in 1939.

What do cat actors say on stage?

Tabby or not tabby!

Tabby Cats at-a-glance

Size: Females are all sizes, males are generally medium to large
Tendency toward obesity (on a scale of 1 - 10): 6
Sex: Both flavors
Color combos: Gray with black stripes, tawny brown with black stripes, silver with black stripes
Peculiar traits: Steady, dog-like, usually gregarious
Craziness (on a scale of 1 - 10): 5
Famous tabbies: Thomasina, R. Kliban's Meatloaf cats

Tabby Cats: the Everyman of the Feline World

They're affable, steady, and devoted. Some people think tabbies are nothing special, but, then, why mess with a classic? They are like the Volvos of the cat world—slightly shapeless in appearance, not exciting, and yet—comforting, and safe. The cartoonist R. Kliban called them Meatloaf cats.

Tabbies are a great type of cat for dog-lovers because they often greet you at the door or beg for attention, and have been known to enjoy taking walks. If you want your party guests to meet your cat, choose a tabby. They'll always mingle.

They're also good outdoors, and fairly athletic. A female tabby will do well in a multi-cat household, but the males like to play the alpha. Still, most tabbies, male or female, take things in stride. They are not very neurotic. My friends Jay and Dorothy have an older tabby, Leo, who has lost the use of his back legs but never seems to notice. He just walks around dragging them, happy as ever. My college roommate had a tabby, Otis, who was like a Buddhist monk, serene and quiet. Otis moved with my friend over and over and never seemed to mind adjusting to a new apartment. Maybe that's why she gave him a name associated with an elevator company.

My friend Anne's young tabby, Hilda, is the queen of the household. She's named after a famous abbess from the Middle Ages. Her adoptive brother, Henry, a large, orange cat, is totally cowed by her, and will even allow himself to be pushed out of a sunny spot by his much younger sister. "She has confidence that she is the most important thing around," says Anne. "She is a real princess."

Tabbies I Have Known and Loved

Cosmo: the Lover-Boy Cat

Cosmo arrived on cue, it seemed, only a few days after our beloved tail-less black cat Shlomo had died (see black cat chapter). Cosmo was young and had a bum foot, but it took him no time at all to slip right into the alpha cat role. He's a classic tabby with beautiful black stripes on a mottled, gray-brown background. He's a magnificent cat with one slight flaw: extreme neediness.

Cosmo seems to require at least four to five hours of petting a day. When he was young he was very skittish. He had been around for weeks, and yet he hadn't let me even really touch him. One night I was writing late in my office, after midnight, and I heard a loud wail. Cosmo came running up and climbed onto my lap, scaled my torso, and put his front legs around my neck. Then he nibbled at my neck and began purring. After that we were the best of friends. Inseparable, actually. He sat with me all day long in my office as I worked. I began to get ideas about writing a novel in which my cat would miraculously transform himself into my lover.

Cosmo is a toilet bowl cat. He *never* drinks from a regular bowl of water. He has scared me once or twice in the middle of the night when I've wandered off to the bathroom and put my rear end down on a furry thing.

Cosmo is one of the only cats I've owned who loves a party. He hangs around and visits every guest. At one gathering he even disgraced himself by jumping into a guest's lap that was already occupied by a very full dinner plate.

The saddest day in Cosmo's life was the day we got a dog. He's never really adjusted—his feline pride has been hurt. Whereas before he was the life of the party, now he confines his sociability to those times when the dog is on another floor of the house, or out altogether. I was surprised—I thought he would be more courageous.

Daisy: the Wallflower

I'm officially classifying Daisy as a tabby, because she has marvelous stripes on a tawny background. And yet . . . she also has a bit of orange in her, and a white bib, so I suppose you could call her a calico. She's certainly timid and nuts enough. But she also has a steady disposition, like a tabby, and when she stands next to her brother Cosmo, a classic, they look very much alike. But Daisy is only about a third of Cosmo's size—I think of her as positively dainty. If she were a female human, she would be one of those annoying size-two girls.

Daisy was a feral cat who showed up pregnant in our backyard. Soon afterwards she gave birth to Sheba, a kitten I placed with a family in North Philadelphia. We were able to lure Sheba away from her mother when she was about six weeks old. But we were never able to catch Daisy.

The weather got colder, and I began to worry. I bought a little wooden doghouse with an asphalt-shingle roof, and put a bunch of blankets and towels in it. Daisy's fur got thick and lush as she lived out the winter in her outdoor condo. She didn't deign to enter our house for another year, and even then the prospect of, say, walking across the living room, terrified her. She never looked at us, and she kept her tail down. She never came within touching distance. This went on for another two years!

Gradually, Daisy became more civilized. "Why Daisy, you're almost a pet," said my husband one day after she had been with us for over five years. And it's true. She will now come and beg a bit for affection. She won't tolerate being picked up, but she does like to be scratched behind the ears. She gets along with every other cat in the household—a real female tabby trait. She's still the same tiny, compact cat she was when she first arrived.

But you can tell that Daisy started out wild, because she doesn't totally trust us, even after all these years. She'll sit in a chair or in front of the fire and appear relaxed, but she never really is. She is always on the lookout for disaster. A sudden noise will send her scurrying for cover. If I have forgotten my keys and run across the room to retrieve them, Daisy will panic and disappear. She has trouble at doorways, I think because she believes we will close the door on her. And any rustling of a paper or plastic bag strikes terror into her heart.

"Yeah, you're right, Daisy," my husband will say in disgust. "We've fed you and taken care of you for eight years, but today is the day we've decided to kill you!"

Great Names for Tabbies

Note: Tabbies are the most dog-like of cats and can handle a hearty name. A lot of people name them after dead relatives, too.

Daisy	Fred
Otis	Gertrude
Hugo	Alice
Cosmo	Mabel
Leo	Hilda
Zeppo	Hermione
LeRoy	Enid
Rover	Bernice
Gummo	Ralph

Alley Cat Dreams

Cats and people have always enjoyed a mystical relationship, and over the centuries people have analyzed their dreams of different color cats, looking for meaning. Below, a survey of European cat dream interpretation!

To dream of a black cat is lucky.

To dream of a tortoiseshell cat means luck in love.

To dream of a ginger cat means luck in money and business.

To dream of a white cat means luck in creativity, spiritual matters, divination, and spellcraft.

To dream of a black-and-white cat means luck with children; it may also mean the birth of a child.

To dream of a tabby cat means luck for the home and all who live there.

To dream of a gray cat means to be guided by your dreams.

To dream of a calico or multicolored cat means luck with new friends and old ones.

Black-and-White Cats

Sufferin' succotash!
—Sylvester, the black-and-white Warner Brothers cartoon star

Black-and-White Cats at-a-glance

Size: Small to medium
Tendency toward obesity (on a scale of 1 - 10): 4
Sex: Both flavors
Color combos: Black and white in many eccentric patterns
Peculiar traits: Good lap cats, great purring capabilities, content to laze in front of the fire, very sweet temperaments, affectionate and easygoing
Craziness (on a scale of 1 - 10): 5
Famous black-and-white cats: Sylvester of Warner Brothers fame, Socks, the First Cat

Black-and-White Cats

I've always liked black-and-white milkshakes, and those famous black-and-white cookies featured on *Seinfeld* episodes.

"They are just wonderful affectionate cats," says my friend Gretchen who owns the very sociable Casper. It's as if the craziness of both black cats and white cats cancels out in the combination, leaving only sweet felines with very big hearts.

The shape of the average black-and-white alley cat is also appealing—they're often a little more rounded than others, with big friendly faces, and interesting markings. I even knew a black-and-white dude called, originally, Hitler because of his little black mustache. He was evidence that you should always try out offensive pet names by screaming them at the top of your lungs several times. My friends were too embarrassed to scream "Hitler," out the back door, and the cat got a friendlier name right away.

Black-and-Whites I Have Known and Loved

Midgie and Minnie: Sisters Forever

God bless Midgie. My husband still cries about her, fifteen years after her death. She was only with us for two years, but she had the most delightful classic black-and-white cat personality.

Midgie was acquired of necessity—our cat Sonny had died, and I was inconsolable. In fact, I didn't think I ever wanted to own a cat again. "I'll switch to dogs," I told my friends, "or ferrets." But my husband thought differently, and he was sick of my sobbing, so on his way home from work one day he spied Midgie, a tiny skunk-like wisp of a thing, in a pet shop. I cried when he produced her from a cardboard box and put her in the palm of my hand.

Yet it soon became clear why Midgie was so small: she was dying. She refused to eat, and by the next morning seemed unable to move. We took her to our vet, where I cried as soon as I entered the examining room. Sonny's death was too recent, and I feared having to go through a kitty tragedy all over again. My vet was furious—the pet shop was unscrupulous, she said, discovering that Midgie was riddled with worms and dangerously dehydrated. It was nip and tuck for a while, but the little oreo cat survived and soon took over the household. One of her most spectacular "tricks" as a kitten was to run about with poop streaming out of her backside, a side effect of her worm-based intestinal problems. It's a testimony to her personality that we loved her anyway.

But Midgie was lonely, I could tell, and I never again wanted to make the mistake of having an only cat. So my husband went off to the shelter and got another black-and-white girl, Minnie. Poor Minnie was also very young—probably only six or eight weeks old—and someone had tied her to the parking meter outside the shelter. Maybe that's why she was always a little spooky, and lacked the usually gregarious black-and-white demeanor.

Midgie loved Minnie from the start. Together, they were glorious kitten sisters who got into everything. My husband was in law school at the time, and we had a lot of impromptu dinner parties during which Minnie and Midgie were the main entertainment. They would dash up and down our metal spiral staircase, pausing sometimes to pounce on a piece of chicken on the table. On one memorable occasion Minnie got stuck under a paper bag and ran around and around the room, looking like a sack with furry legs.

At night we had a set routine. We would always hope that they would settle down, but just when we were falling asleep, the kittens would start pouncing on our faces and playing with our hair. That's when they had to be locked into what we called the clubhouse—an old bathroom of ours with a claw tub and ratty shower curtain. In that tiny space, they would play the night away, climbing up and down the tub and hanging from the curtain. In their clubhouse, the girls would eat anything, even food they refused to eat in the kitchen. They were very nocturnal, so by the time we retrieved them in the morning they were ready to doze all day.

Instead of being mousers, Minnie and Midgie went after insects. The year they were kittens we had an amazing infestation of huge grasshoppers that the girls would drag in through the open window. I tried in vain to save as many of the huge, green bugs as I could, but there were always, at any time, at least three of them being tortured in some hidden location. That summer we went away for a while, and our nearsighted cat sitter said that everything had gone well, but she could never understand why there were so many "plastic pellets" on the floor. Were they cat toys, she asked? After she left we stooped down to pick up the pellets and discovered that they were actually scores of grasshopper "fuselages," only the torsos devoid of legs. Blech!

Every cat-lover has an ailurophobic friend, one who just can't bear the thought of a cat ever touching her. One of my friends, Susan, was sitting in a chair near an open window one fine fall evening when Midgie appeared and jumped onto her shoulder. That would have been bad enough, but Midgie had a huge squirming moth in her mouth. BZZZZ, it went, as the sides of Midgie's maw buckled and strained. Susan jumped straight out of her chair, got her coat, and didn't come back to visit for several years.

Minnie loved to help me make the bed, and Midgie would join in. Together they would run around under the sheets as I tried to tuck in the corners. The game was to trap them and then listen to them purr until they snapped and started clawing at me through the covers. Then they would butt their heads up against the tucked-in blankets and eventually make it out of the bed.

I have many, many pictures of the cute black-and-white sisters asleep on a chair, in boxes, and next to my newborn baby. Alas, though, Midgie died a young and mysterious death of internal injuries from, we think, a fall. We found her crouching under a tomato plant, deeply in shock. Our vet tried hard to save her, but she didn't make it through the operation. I know it sounds far-fetched and romantic, but her sister Minnie never really got over her death. For weeks afterwards she wandered around the house crying piteously. Even though she lived to a ripe old age, Minnie became withdrawn and never really played much again. Still, the images of those two cute kittens are some of the best cat memories I have.

Felix: the Traveler

Okay, so I know that's a corny name for a cat. But he was one of my most beloved. Felix was a country boy. I found him after his feral mother had been shot by some obnoxious hunters who bragged about using wild cats for target practice. I don't know if they killed any of Felix's brothers and sisters. At the time I lived in a strange chicken coop on forty acres near a stream. When the apartment next to me was vacant, the landlord would rent it out to noisy, drunken deer hunters.

I thought Felix was retarded during the first day after I found him. He could barely lift his head, and was unable to focus on anything in front of him, even food. But once I started forcing tiny bits of tuna down his throat, he came alive. He had merely been starved—I don't know how long he'd been without his mother's milk.

Felix was predominately white, with several black spots and a black widow's peak on his head. He was one of those rare cats who really liked to take walks. He would follow me all around the woods near the chicken coop, and along the stream. We picked blackberries together—really, it was like a scene from an idyllic movie. Unlike many black-and-white cats, he wasn't much of a lap cat (a friend has a theory that the more white in the background, the less chummy the cat is!), but he was very affectionate and devoted, and a great walking companion.

When I moved to the city, my parents offered to take Felix, knowing that he would be miserable cooped up in a tiny, studio apartment. He had only been at my folks' for a few weeks when I got a distressed call: my boy was missing. My parents had looked everywhere, frantically, but there was no sign of him. After a few weeks I gave up hope that he would ever return.

But two months later I got a jubilant phone call from my father at seven A.M. "He's back!" he screamed into the phone, knowing that I would know exactly whom he was talking about. Felix had arrived back about half his former weight, dirty, and disheveled. Where had he been? My mom theorized that he had jumped into someone's car and had been driven miles away yet managed to make it back. I thought maybe he had set out on foot to find my old place, which was about eight miles away. These are the situations when a cat's inability to talk can get very frustrating.

Felix lived out his days in the country, becoming an expert hunter. I once saw him snap the neck of a rabbit nearly his own size. After my father died, my mother moved to a small house on a river, taking Felix with her. He adapted well, catching snakes and leaving them as trophies on my mother's patio. He never disappeared again.

Great Names for Black-and-White Cats

Domino

Panda

Oreo

Keys

Newsy (for news print)

Skunk

Pepe Le Pew

Tuxedo, or Tux

Non Pareil

The Cats of Many Colors

Many times well-meaning people will ask: which color cat is your favorite?

How can I answer? That's a little like being asked to compare, say, flowers or food. As much as I like daisies, would I really never want a bouquet of roses? And I adore pumpkin pie, but I also like an occasional slice of apple. Different colored cats fit different moods, though. It's a pity that we can't rotate cats as we do furnishings and draperies—tabbies and gray cats atop our velvet sofas for winter, white cats and orange cats perched on wicker settees for the spring and summer months. Of course, cat décor would only be possible if we had cat storage warehouses somewhere to handle the rotation, and we all know how crazy that would be. So I guess we are stuck with the luck of the draw—whatever color cat comes my way, I tend to get attached to it.

Yet I've definitely experienced runs of certain color cats. In my family for a while, all our cats had white backgrounds, and so my mother tended to pick lighter fabrics to camouflage the several hundred pounds of fur they shed onto the furniture each month. When I left home, I took an old couch and chair with me, only to end up with two gray cats whose sheddings were not color-coordinated with the upholstery. Gradually, as I began to buy new furniture, I gravitated towards darker pieces, only to find that stray cats with white background hairs were showing up at my house. I've given up now and tend to see cat hair of all colors as an attractive textural garnish to any fabric.

My father, who refused to learn cat names and referred to every one of our cats as "Hey you with the fuzzy face," often raised a more significant question about felines having nothing to do with color preference: "Why do we own these little buggers, anyway?" he would rail when some cat's shenanigans resulted in a broken vase or an unrolled expanse of toilet paper straggling throughout the house. "They're no bigger than chickens. We *eat* chickens!" My mother was less vocal, yet more of a disciplinarian. When our half dozen cats would begin to get on her nerves, she would give them what we called "the cellar treatment." They would be unceremoniously dumped down the basement steps to run around to their heart's content, where no one had to worry about them batting the defrosting pork chops off the kitchen counters.

In the end, I think we own cats because we have to. Cat ownership is not as much of a conscious choice as we think. Cats are rather Zen-like creatures. When they land some place, it usually seems as if it's for a reason. The common joke among all of us hapless cat fanciers is, "I don't own my cat. My cat owns me." True. For as much as our cats' hair gets into our noses, their souls get under our skins. They become living, purring extensions of us, and their calming effects were around long before Prozac or Valium.

Whenever I think of the lengths I've gone to for my cats—the special food, the thousands of dollars in vet's visits, the little presents of hairballs, dead mice, or inappropriate poops they leave now and then—I tell myself that if the prophet Mohammed could be a slave to a cat, then how bad is it that I adore my felines?

As the story goes, Mohammed's cat fell asleep next to the prophet on part of his robe. Mohammed had to get up, but instead of disturbing his kitty, he actually cut part of his robe off so that she could slumber in comfort. I understand the holy man's actions—which one of us wouldn't give the shirt off his or her back for the best cat in the world?

About the Author

Cathy Crimmins, the author of eighteen books, has owned over forty cats in her lifetime and has been a friend to many more. Since her grandfather ended his days as the crazy owner of thirty stray kitties, Crimmins is destined to become a cat lady in her old age. She lives in Philadelphia, where the alleys feature felines galore.